Games & Puzzles

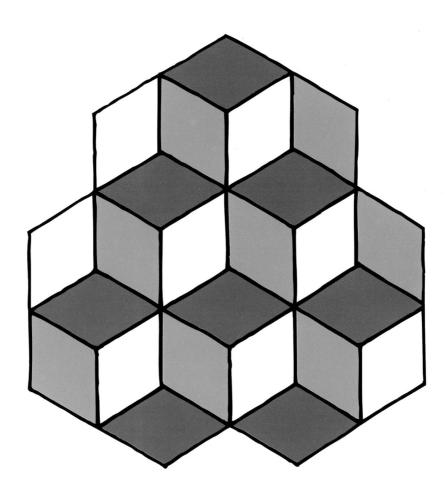

David Kirkby

RIGBY
INTERACTIVE
LIBRARY

© 1996 Rigby Education
Published by Rigby Interactive Library,
an imprint of Rigby Education,
division of Reed Elsevier, Inc.
500 Coventry Lane,
Crystal Lake, IL 60014

Designed by The Point
Cover design by Pinpoint Design
Printed in the United States of America

00 99 98 97 96
10 9 8 7 6 5 4 3 2 1

Library of Congress Cataloging-in-Publication Data
Kirkby, David, 1943–
    Games & puzzles / David Kirkby.
       p. cm. — (Math live)
    Includes index.
    Summary: Presents simple mathematical concepts in games or puzzles
using dominoes, coins, straws, mazes, palindromes, solitaire, tangrams,
and others.
    ISBN 1-57572-044-2 (lib. bdg.)
    1. Mathematical recreations—Juvenile literature.
[1. Mathematical recreations.] I. Title. II. Series: Kirkby,
David, 1943– Math live.
QA95.K486 1996
793.7'4—dc20   95–36240
                          CIP
                          AC

Acknowledgments
The author and publisher wish to acknowledge, with thanks, the following photographic sources:
Skyscan Balloon Photography, p. 18; 1995 MC Escher/Cordon Art Baarn Holland; Trevor Clifford, p. 40.

The publisher would also like to thank the following for the kind loan of equipment:
NES Arnold Ltd; Polydron International Ltd.

Note to the Reader
In this book some words are printed in **bold** type. This indicates that the word
is listed in the glossary on page 44. The glossary gives a brief explanation of
words that may be new to you.

CONTENTS

15.95

① DOMINOES

Dominoes are used for many different games and puzzles. A full set contains 28 dominoes, ranging from a double blank to a double six. In many domino games you have to join one domino to another domino on the table, so that they meet with a matching number of dots.

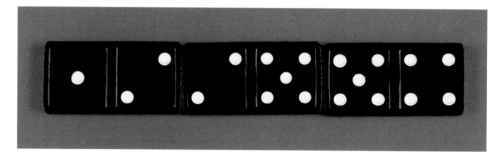

It is possible to make a chain of dominoes using all 28, so that they meet with matching numbers of dots. It is also possible to make matching shapes. Below are two different domino squares.

DOMINO RECTANGLES

Using a set of dominoes, can you make domino rectangles, meeting with matching numbers of dots to fit these arrangements?

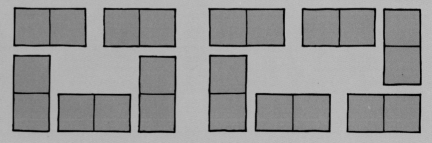

If the dominoes with a blank are removed, and the others are sorted into columns according to their total numbers of dots, then a **symmetrical pattern** is created. This shows that there are three dominoes that have a total of six dots, three that have seven, and three that have eight.

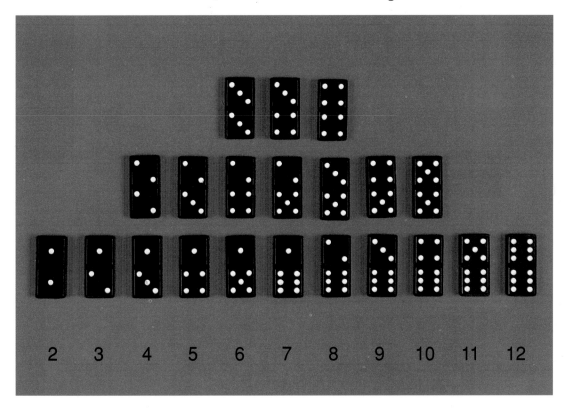

DOMINO ADDITIONS

It is possible to arrange three dominoes to make addition sums like the one shown.

Can you make some more?

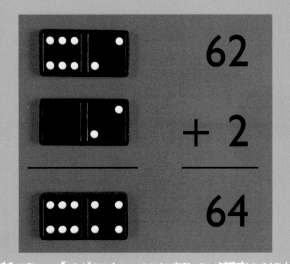

$$62$$
$$+ 2$$
$$\overline{64}$$

COINS

Many puzzles use coins. One reason is that they are easy to slide, although in some cases counters can be used as a replacement. Another reason is that each face is different (head or tail).

COIN ILLUSION

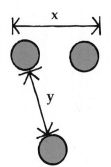

Try this illusion with coins to test your skills of observation. Place three coins in a row. Slide the middle coin until you think that the distances x and y are the same. Then use a ruler to see if you are correct.

FLIPPING COINS

Start by placing nine coins like this, so that there are eight heads and one tail. A move consists of turning over all three coins in any row, column, or **diagonal**. Can you finish with all nine coins heads up? It can be done in 5 moves.

COIN SWAP

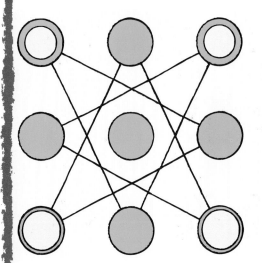

Place two small and two large coins like this. Slide a coin along a line to an empty spot. Can you swap the positions of the small and large coins? How many moves does it take?

Other coin puzzles use the value of the coins themselves.

COIN SHARING

In this puzzle you need the coins shown here.

The money must be shared equally between two people.
Can you find three different ways of sharing the money?

COIN GAME

This is a game for two players to practice skills of addition. You need one coin. The first player places the coin on any spot and says the number. The second player slides the coin along a line to another spot, and says the running total by adding on the new number on the spot. Continue taking turns sliding the coin and saying the running total. The first player to say a total that is more than 30 loses.

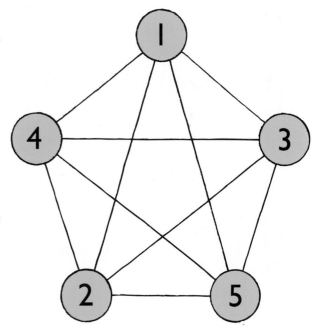

3 STRAWS

There are many different types of straw puzzles. Many involve making shapes.

STRAW SQUARES

Make this shape with short straws.
- Remove 2 straws to leave 5 squares.
- Remove 3 straws to leave 4 squares.
- Remove 4 straws to leave 3 squares.
- Remove 4 straws to leave 4 squares.
- Remove 5 straws to leave 4 squares.

STRAW DIVISION

This type of puzzle is based on adding more straws. Make this shape with short straws. Add 8 more short straws so that the shape can be divided into four equal parts.

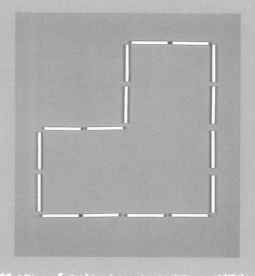

STRAW EQUATIONS

These puzzles use short straws to make Roman numerals.

Make these equations with straws. In each case move one straw to a new position to make each equation correct.

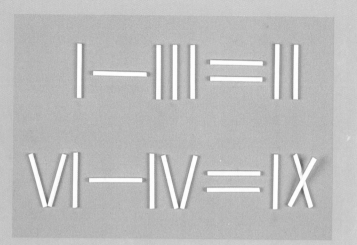

NIM

This is a game for two players.
• Take a handful of straws and make two piles.
• Take turns removing any number of straws from one pile.
• The winner is the player who takes the last straw.

4 CARDS

There are hundreds of different games, tricks, and puzzles based on a deck of cards.

DOUBLES

Start by removing the picture cards. The aces represent the digit 1. The idea is to make pairs of numbers so that one number is double the other.

Here are some examples.

Find your own deck of cards and see how many "doubles" you can make.
How many "triples" can you make?

TRIPLETS

Use the ace to ten of spades only.
Find a set of three cards that total 12: for example, 2, 3, and 7.
It is possible to find seven different triplets altogether.
Can you find them?

There are nine different triplets that have a total of 15.
Can you find them?
You must make sure that they are all different!

GUESS MY CARD

This is a game for two players.
- Player A chooses a card, looks at it, then hides it.
- Player B has to guess what the card is by asking a series of questions.
- Player A may only answer "Yes" or "No."
- Count how many questions are needed before the card is guessed correctly. The winner is the player who needed fewer guesses. Swap roles, so that B has to guess A's card.

SIXTEEN CARD PUZZLE

This is a card puzzle that seems easy to start with, but then becomes more and more difficult.

You need to take the four aces, the four twos, the four threes, and the four fours from the deck. The sixteen cards have to be arranged in a 4 x 4 pattern.
- Can you arrange the cards so that the same number does not appear in any row?
- Now can you arrange them so that the same number does not appear in any row or column?
- Finally, can you arrange them so that the same number does not appear in any row, column, or diagonal?

NUMBER CUBES

Number cubes have been used in games for hundreds of years. This is because when they are rolled, each face has an equal chance of facing upward, and so winning or losing depends on chance. The most common type of number cube has dots on each of the six faces, but all sorts of different cubes are available.

MAKE A SIX-FACED NUMBER CUBE

You need thin cardboard.
- Use a ruler to draw this net of six squares.
- The edges of the squares should be 3 inches.
- Add tabs, cut it out, and crease along the lines.
- Draw the dots on the squares.
- Fold it and glue the tabs to make your number cube.

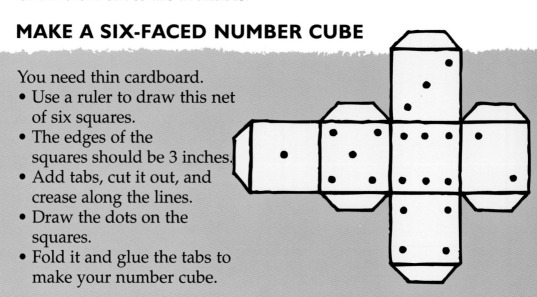

MISSING DOTS

The opposite faces of a six-faced number cube with dots always total 7. These are nets of cubes with dots missing from some of the faces. Can you figure out how many dots there should be on each missing face?

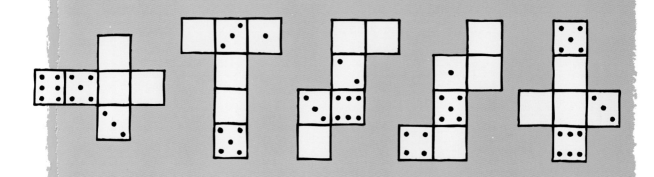

IMPOSSIBLE NUMBER CUBES

Some of these drawings are different views of the same number cubes. Some are impossible. Which are the impossible ones? Use a number cube to help you.

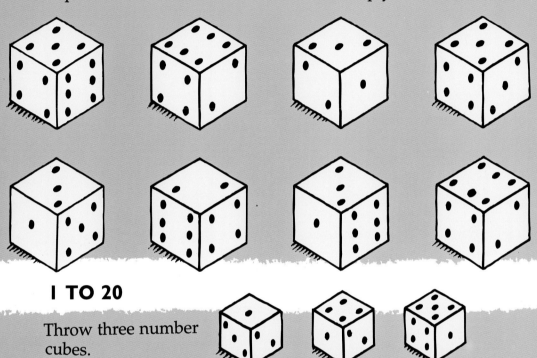

I TO 20

Throw three number cubes.

Then see how many of the numbers 1 to 20 you can make using the three numbers on your cube. Each cube number can only be used once each time. But you don't have to use all three numbers. Here are the answers using 1, 4, 5.

① = 5 - 4	⑥ = 5 + 1	⑪ = 15 - 4	⑯ = (5 - 1) × 4
② = 5 + 1 - 4	⑦ =	⑫ =	⑰ =
③ = 4 - 1	⑧ = 5 + 4 - 1	⑬ =	⑱ =
④ = 5 - 1	⑨ = 5 + 4	⑭ =	⑲ = 15 + 4
⑤ = 4 + 1	⑩ = 5 + 4 + 1	⑮ = (4 - 1) × 5	⑳ = 4 × 5

NUMBER CUBE GAMES

You can play many different games with number cubes. The ones described here require number skills. In the first game, you need to be able to keep a running total of your score as you throw the cube.

THREE LIVES

This is a game for two or more players.
You need one number cube. On your turn, throw the cube, and keep a running total of your score. Each time you throw a 1 you lose a "life." When you have lost three "lives," your turn ends and you record your score. The winner of the round is the player with the highest total.

NUMBER STRIPS

This is a game for two players.
You need at least four number cubes. On your turn, throw as many cubes as you like at once. Find their total, and place a counter on the same number on the board. If this is not possible, or the number is already covered by a counter, then do nothing. The winner is the first player to get *either* 5 counters in any one strip, *or* 4 counters on each of two strips.

1	2	3	4	5	6
7	8	9	10	11	12
13	14	15	16	17	18
19	20	21	22	23	24

FIVE HOME

This is a game for one person.
You need one number cube and five counters. Place one counter on each of the five starting spots.

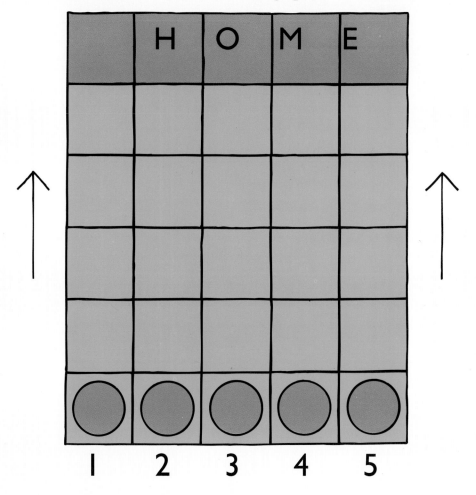

Throw the cube and move counters to match the number you rolled. For example, if you throw a 5 then you can move:
• the 5-counter one space forward
• or the 1-counter five spaces forward
• or the 4-counter and 1-counter each one space forward
• or the 3-counter and 2-counter each one space forward.
Each time you throw the cube, make a tally mark on a sheet of paper, to record how many throws you have had.
Can you get all the counters "home" in fewer than 25 throws?

BOARD GAMES

It is easy to make a board like the one on the opposite page. Use it to play different games. These games are all for two players. You each need a set of counters of your own color.

SUM AND DIFFERENCE

You need two number cubes.
Take turns throwing both cubes and placing a counter on a number on the board to match: *either* the sum of the two numbers on the cube *or* the difference between the two numbers. If you cannot place a counter, do nothing.
The winner is the first player to have four counters in any one straight line (horizontal, vertical, or diagonal).

MULTIPLICATION

You need a pack of cards with the picture cards removed. Shuffle them and place them face down in a pile. Take turns picking up the top two cards and multiplying the card numbers together. Place a counter on a number that matches the units digits of the answer.

For example, if the cards are 7 and 4, multiply them to make 28, and place a counter on 8. The winner is the first player to have three counters in any one straight line.

SUBTRACTION

You need two number cubes.
Take turns throwing either one or two cubes. If you throw two, find the total. Subtract your score from 12, then place a counter on a matching number. When all the numbers have been covered, the winner is the player who has placed the most counters.

2	5	4	8	9
0	3	6	2	5
8	6	1	7	4
9	5	0	8	3
1	7	4	6	7

TO DO

Make your own board by writing numbers in a grid.
Invent some rules for a game and play it.

MAZES

One of the most famous mazes in the world is at Hampton Court Palace in England. The maze is built with high hedges.

MAZE CHALLENGE

Can you find the quickest way through the maze?

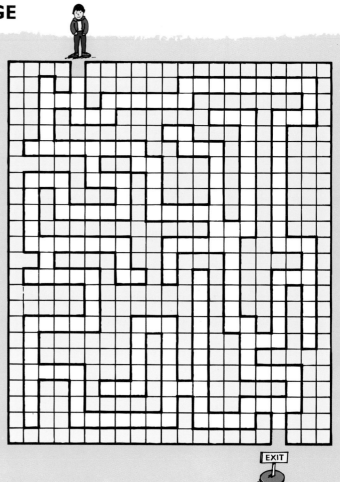

EXIT

MAKE YOUR OWN MAZE

Start with an 8 x 8 square grid.
1 Place the hidden treasure in one of the squares. Draw a
   route from the treasure to the outside (the entrance).
2 Draw some fences.
3 Then copy the drawing again, without showing the route.

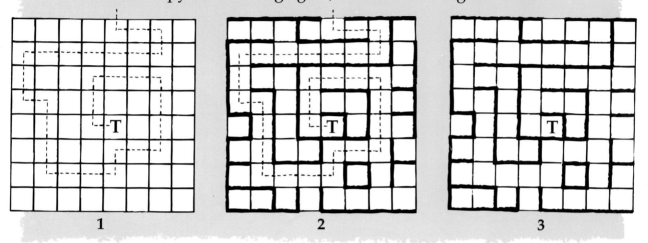

1   2   3

NUMBER MAZE

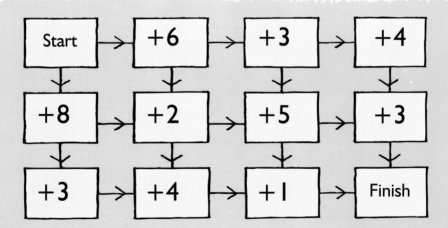

Find a pathway from Start to Finish. You must follow
the direction of the arrows.
• Can you find a pathway that totals 15?
• Which pathway gives the largest total? Which pathway
  is the smallest total?
• How many different totals are possible?

19

9 TARGETS

Many sports and games are based on aiming at a target and scoring points.

TARGET TOTALS

Look at these three targets.
Suppose you can have three shots at each target.
What is the largest possible total score for each target?

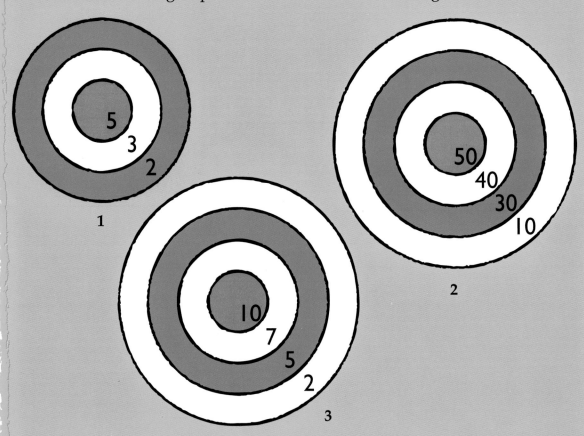

What are all the different possible scores with three shots at target 1? How many different scores are possible for targets 2 and 3?

TO DO

Invent your own target and points for each ring.
Investigate all the different possible scores with three shots.

Another well known target game is darts.

A Darts thrown into the outer ring of the
board count double the score.
B Darts in the inner ring count triple the score.
C Darts in the inner bullseye count 50 points.
D Darts in the outer bullseye count 25 points.

There are many different variations of
the game of darts. In most of these,
players are allowed to throw three
darts at each turn.

*What score is made with the darts in
the photograph?*

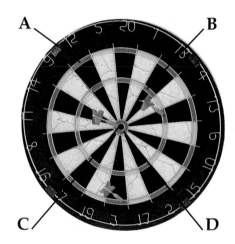

NUMBER CUBE DARTS

This is a game for two or more players.
You need one cube numbered 1 to 6.
Make another by writing S, S, D, D, T, T on the faces of a
cube (S stands for "single," D for "double," and T for
"triple").
Take turns throwing both cubes, and find your score.
If you throw 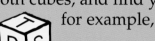 for example,

then your score is triple 4, or 12.
Play ten rounds each, keeping score on a scoresheet.
The winner is the player with the largest total score.

VARIATION

Now write 5, 6, 7, 8, 9, 10 on the faces of another cube,
and use this instead of the cube numbered 1 to 6.

NUMBER TRICKS

Here are some number tricks to amaze your friends. They are devised so that they always produce the same answer. The first trick, for example, always gives the answer 6.

ANSWER 6

Start by secretly writing a large number 6 on a piece of paper, then placing it in an envelope and sealing it. Give the sealed envelope to a friend.

Then ask your friend to:
• write down any number
• double it
• add 20
• subtract 8
• divide by 2
• subtract your original number
• open the envelope!

Here are two more envelope tricks to try.

ANSWER 23

Seal the number 23 in an envelope. Tell your friend to:
• write down any three-digit number
• add 25
• multiply by 2
• subtract 4
• divide by 2
• subtract your original number
• open the envelope!

ANSWER 3

Seal the number 3 in an envelope. Tell your friend to:
• write down any number
• triple it
• add 12
• double it
• subtract 6
• divide by 2
• divide by 3
• subtract your original number
• open the envelope!

Try this trick on a friend. You need 3 number cubes.

ADD 7

Ask a friend to throw the three cubes, unseen by you.
Tell your friend to:
• find the total
• turn one of the cubes upside down and add the number
  on the bottom
• roll this cube again and add whatever number appears
• write down the total score in secret.
You can now look at the three cubes
and say what the total is.

Secret

Add seven to the total of the three
cubes you can see.

Here are two more number cube tricks to try.

SUBTRACT 150

Ask a friend to throw three
cubes, unseen by you.
Tell your friend to:
• double the number on the
  first cube
• add 3
• multiply by 5
• add the second cube number
• multiply by 10
• add the third cube number
• tell you the result.
You can now say the scores on
the three cubes.

Secret

Subtract 150 from the result.

SUBTRACT 110

Ask a friend to throw three
cubes, unseen by you.
Tell your friend to:
• multiply the number on the
  first cube by 10
• add 11
• add the second cube number
• multiply by 5
• double it
• add the third cube number
• tell you the result.
You can now say the scores on
the three cubes.

Secret

Subtract 110 from the result.

The digits of your answer are the numbers on the cubes!

11 AGE TRICKS

Here are some tricks to help you find out someone's age.
Try them out on yourself first.

AGE TRICK 1

Ask someone to:

- write down any three-digit number
- rearrange the digits to make another three-digit number
- subtract the smaller number from the larger
- add your age (say, 10)
- tell you the answer.

You can now tell them their age.

$$\begin{array}{r} 613 \\ -136 \\ \hline 477 \\ +10 \\ \hline 487 \end{array}$$

Secret

Add the digits of the answer until you have a single-digit number, for example, $4 + 8 + 7 = 19$, then $1 + 9 = 10$ and $1 + 0 = 1$.
Add this number to a **multiple** of 9 which gives an approximate age: $1 + 9 = 10$.

AGE TRICK 2

Ask someone to:

- write down his or her address number
- double it
- add the number of days in a week
- multiply by 50
- add his or her age (say, 11)
- subtract the number of days in a year (365)
- add 15.

	71
$71 \times 2 =$	142
$142 + 7 =$	149
$149 \times 50 =$	7450
$7450 + 11 =$	7461
$7461 - 365 =$	7096
$7096 + 15 =$	7111

address number / age

You can figure out the address number and age from the answer.

AGE TRICK 3

Ask someone to look at the table and say in which columns his or her age appears. Then add together the top numbers in each of these columns to find the age.

For example, if the age is in the first, third, and fifth columns, then you add together 1, 4, and 16 to give the age of 21.

1	2	4	8	16	32
3	3	5	9	17	33
5	6	6	10	18	34
7	7	7	11	19	35
9	10	12	12	20	36
11	11	13	13	21	37
13	14	14	14	22	38
15	15	15	15	23	39
17	18	20	24	24	40
19	19	21	25	25	41
21	22	22	26	26	42
23	23	23	27	27	43
25	26	28	28	28	44
27	27	29	29	29	45
29	30	30	30	30	46
31	31	31	31	31	47
33	34	36	40	48	48
35	35	37	41	49	49
37	38	38	42	50	50
39	39	39	43	51	51
41	42	44	44	52	52
43	43	45	45	53	53
45	46	46	46	54	54
47	47	47	47	55	55
49	50	52	56	56	56
51	51	53	57	57	57
53	54	54	58	58	58
55	55	55	59	59	59
57	58	60	60	60	60
59	59	61	61	61	61
61	62	62	62	62	62
63	63	63	63	63	63

There are interesting patterns of numbers in some dates.

CENTURY DATES

March 1, 1996, is a century date
  3 – 1 – 96
because 3 + 1 + 96 = 100.
Find some more century dates.

CONSECUTIVE DATES

October 9, 1987, is a consecutive date
  10 – 9 – 87
so is June 7, 1989,
  6 – 7 – 89
Why is March 13, 1993, a strange date?
Can you find more like it?

12 PALINDROMES

A palindrome reads the same forward and backward.
Here are some examples of palindromic words.

level	gag	rotator
deed	noon	civic
did	pip	radar

Palindromic Party Guest list

Mom	Bob
Dad	Lil
Anna	Hannah

Some names are palindromic. *Who else could we invite to the party?*

Some people have spent a long time trying to create palindromic sentences. Here are some examples:

Madam, I'm Adam.
Norma is as selfless as I am, Ron.
Doc note, I dissent, a fast never prevents a fatness, I diet on cod.

Palindromic numbers can be of different sizes.
Here are some palindromic three-digit numbers.

THREE-DIGIT PALINDROMES

There are another 27 palindromic numbers between 100 and 400. Can you find them?

171
222
363

If we start with any two-digit
number, say   26
Reverse the digits and add: 26 + 62 = 88
                                    (a palindrome in 1 step)

Now let's try with 57   57
Reverse the digits and add: 57 + 75 = 132
Reverse these digits and add: 132 + 231 = 363
                                    (a palindrome in 2 steps)

TWO-DIGIT PALINDROMES

Try this with your own two-digit numbers.
Do all two-digit numbers become palindromes?

PALINDROMIC TIMES

The clock time is palindromic.
Find some more palindromic
times.

PALINDROMIC DATES

Some dates are palindromic.
What is the next palindromic date?
1991 was the last palindromic year. What is the next?

OPTICAL ILLUSIONS

An optical illusion is a picture or drawing that seems to play tricks on your eyes.

Study this picture. What is wrong with it?

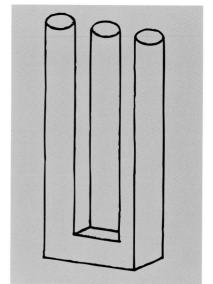

These drawings will test your eyes.

*Which **horizontal line** is longer, A or B?*

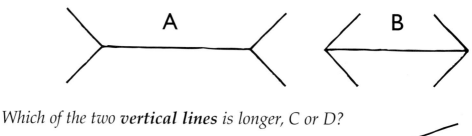

*Which of the two **vertical lines** is longer, C or D?*

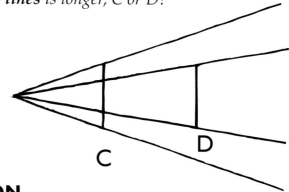

DRAW AN ILLUSION

Draw two straight lines that are the same length, and label them A and B.

Draw some more lines to try to make A and B appear to be of different lengths.

Ask a friend to judge which is the longer.

COUNTING CUBES

Look carefully at this picture, until you can see some cubes. Count them.

Keep looking until you can see a different set of cubes, and count these.

It is possible to count one set of six cubes and one set of seven cubes.

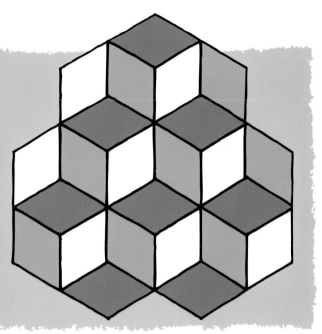

M.C. Escher, a Dutch artist, created many pictures that contained optical illusions. This one, called *Belvedere*, was drawn in 1958.

Study it carefully, and pay particular attention to the ladder.

14 MAGIC SQUARES

A **magic square** is a square grid containing a different number in each square of the grid. It is called "magic" when all of its rows, columns, and diagonals have the same total.

Check that this is a magic square. The total of each row, column, and diagonal is 15.

Its **magic number** is 15.

4	9	2
3	5	7
8	1	6

MAGIC SQUARES

Here are some more magic 3 x 3 squares. What is the magic number in each case?

3	10	5
8	6	4
7	2	9

5	12	7
10	8	6
9	4	11

4	11	6
9	7	5
8	3	10

NON-MAGIC SQUARES

Make nine cards numbered from 1 to 9, and place them in a 3 x 3 square grid. Can you arrange them so that all the rows, columns, and diagonals have different totals?

Albrecht Dürer, a German artist, created this 4 x 4 magic square in 1514. Check that its rows, columns, and diagonals all total 34. This square is extra special because the magic number of 34 appears in the total of the numbers in all four corners, the four central squares, and each of the four corner 2 x 2 squares.

16	3	2	13
5	10	11	8
9	6	7	12
4	15	14	1

What is the magic number of this 5 x 5 magic square?

12	23	9	20	1
19	5	11	22	8
21	7	18	4	15
3	14	25	6	17
10	16	2	13	24

THE FIFTEENS GAME

This is a game for two players.
Use a set of cards numbered from 1 to 9. Draw a 3 x 3 grid on paper, large enough to fit a card onto each square.

One player uses the odd-numbered cards, the other uses the even-numbered cards. Take turns placing a card in a square in the grid.

The winner is the first player to complete a line of three, horizontally, vertically, or diagonally, that totals 15.

It is possible to create other magic shapes. This is a magic triangle. *Check that its magic number is 12.*

MAGIC TRIANGLES

Can you rearrange the numbers from 1 to 6 in this triangle to create a magic triangle with a magic number of 11? Now try for 10 and 9.

15 MULTIPLYING

Some multiplications can be done with fingers and thumbs.

FINGER MULTIPLYING BY 9

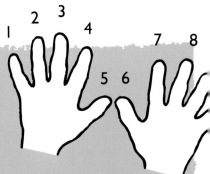

Here is a trick for multiplying a number by 9. Hold out both hands with fingers and thumbs outstretched. Imagine that they are numbered as in the diagram.

To multiply 4 by 9, bend down finger 4.

Count the number of fingers and thumbs that are pointing on either side of this finger. There are 3 on the left (this is the "tens"), and 6 on the right (this is the "units").
So the answer is 36.

Try using this method to multiply 7 by 9, and 3 by 9.

MULTIPLYING BY 11

There is a trick that makes multiplying by 11 easy.
Look at these:

$$17 \times 11 = 187$$
$$23 \times 11 = 253$$
$$35 \times 11 = 385$$
$$42 \times 11 = 462$$

Use a calculator to try some more multiplications by 11.
Can you spot any pattern?

MORE FINGER MULTIPLYING

Here is another trick for using the fingers and thumbs to multiply. Hold out both hands with fingers and thumbs outstretched again. Imagine that they are numbered as in the diagram.

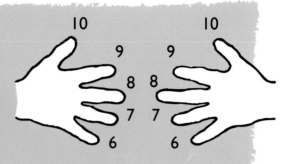

To multiply 7 by 8, touch together finger 7 and finger 8. Count the number of fingers and thumbs that are touching and below. There are 5 (this is the "tens").

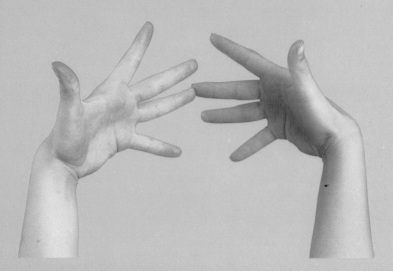

Count the number that are above on each hand. There are 3 and 2. Multiply them together: 3 x 2 = 6 (this is the "units"). So the answer is 56.

Try using this trick to multiply 8 by 6, 9 by 7, and 8 by 8.

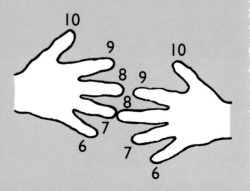

16 CALCULATORS

Some calculator digits can look like letters of the alphabet when turned upside down.

If a large number is keyed into the calculator, which is then turned upside down, it is possible to create a word.

Turn this page upside down to read the word on the calculator display.

CALCULATOR CHALLENGE

Use your calculator to make some words of your own.

CALCULATOR STORIES

Here is a calculator story for you to translate:

This is a story about _ _ _ _	4121 + 1542
which are surrounded by _ _ _ _ _ _	115469 x 5
They come in different _ _ _ _ _	5 x 10643
I prefer the _ _ _ ones	103 x 6
You can eat them raw or you can _ _ _ _ them,	1777 x 4
but many people fry them in _ _ _.	10 x 71

Create your own calculator story.

BACK TO THE START

Here are two tricks with a calculator. Check the calculations using your calculator.

Choose a three-digit number, say 568 568
Repeat the digits to make a six-digit number 568568
Divide by 13 43736
(it divides exactly!)
Divide by 11 3976
(it divides exactly again!)
Divide by 7 568
(it divides exactly yet again!)
Notice that the result is the same three-digit number we started with. *Try this with other three-digit numbers. Do you always finish up with the number you started with?*

Choose any two-digit number, say 73 73
Repeat the digits with zero in between 73073
Divide by 13 5621
(it divides exactly!)
Divide by 11 511
(it divides exactly again!)
Divide by 7 73
(it divides exactly yet again!)
Again the result is the same three-digit number we started with. *Does this always work?*

FIND THE NUMBERS

Try these three calculator tricks on a friend

- Write down any number less than 10.
- Multiply by 7.
- Multiply by 16.
- Subtract the chosen number.
- Divide by 3.
- Divide by 37.

1

- Write down any number less than 10.
- Multiply by 5291.
- Multiply by 21.

2

- Write down any 2 two-digit numbers.
- Multiply one of them by 5.
- Add 36.
- Multiply by 20.
- Add the other two-digit number.
- Subtract 720.

3

⑰ SOLITAIRE

Solitaire is a game that was invented by a Frenchman in the eighteenth century, while he was in prison.

It is a one-person game.

The board has 33 holes. 32 pieces are placed, one in each of the holes except for the center hole.

A move consists of jumping one piece over one other to an empty hole beyond. The piece that has been jumped over is removed. Moves can only be horizontal or vertical, not diagonal.

The object is to continue jumping and removing pieces until you have just one piece left.

The perfect finish is for this one last piece to be in the center of the board.

The game has lots of variations using different board shapes and different numbers of pieces.

This game of Solitaire is played on a board shaped like a **pentagram**.

The rules are the same, and the objective is to finish up with just one counter left.

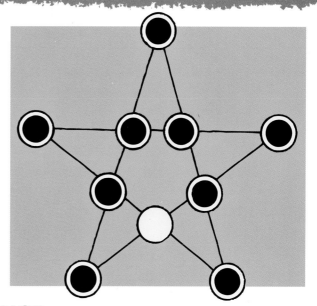

PENTAGRAM CHALLENGE

Can you do it?
Try on this board using counters.

This is a variation of the Solitaire game which includes scoring. The board is triangular.

Start by placing nine counters on any of these ten spots, leaving one clear.

Jump one counter over another, removing the jumped counter in the usual way.

Continue jumping and removing counters until you can jump no more, then add up the numbers on the empty spots. This is your score.

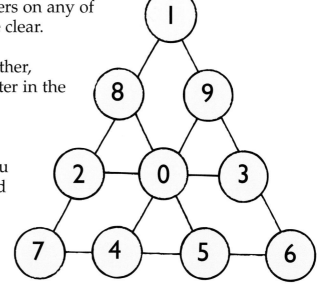

TO DO

Play several games to see what is your best score.

GAMES WITH LINES

There are many games in which the aim is to complete a straight line of pieces. In some games, you have to try not to make a straight line.

This drawing shows that it is possible to place six counters on this 3 x 3 grid without completing a straight line of three counters.

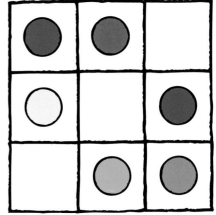

CHALLENGE

Place counters on this 4 x 4 grid. What is the highest number of counters you can place without having three counters in any one straight line?

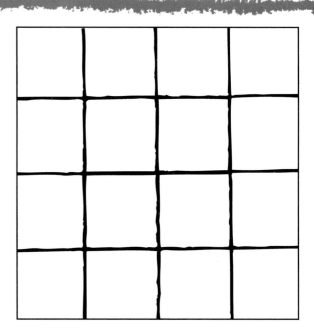

THREE-IN-A-LINE

This is a game for two players.
You each need some counters, of your own color.
Take turns to place a counter on the 4 x 4 grid.
Game 1. The first player to have three counters in a straight line loses.
Game 2. The first player to have three counters in a straight line wins.

Another well known game in which you aim to complete a straight line is the game called "Tick-Tack-Toe."

In this game, you start by drawing two pairs of crossed lines, to make nine spaces. Two players ("O" and "X") take turns writing their marks in one of the nine spaces. The winner is the player who first completes a straight line of three marks.

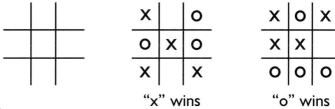

"x" wins   "o" wins

TICK-TACK-TOE

Play Tick-Tack-Toe with a friend.
Then try these three variations of the game.

Variation 1. The first to get three in a line loses.

Variation 2. On your turn you may place your mark in as many spaces as you like, provided that they are all in the same straight line. They do not have to be next to each other. The player who makes the last mark or marks wins.

Variation 3. Play on a 4 x 4 grid. The first player to make a straight line of three in any direction wins.

THREE-DIMENSIONAL TICK-TACK-TOE

A much more difficult variation of the game is to play it in three dimensions. Players place colored marbles, instead of marking Os and Xs. There are 27 different positions in this game. The aim is still to make a straight line of three marbles, but they can be in any direction—horizontally, vertically, or diagonally, and on any level.

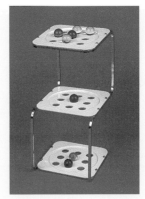

FROGS

"Frogs" is one of many games and puzzles that are based on swapping the positions of two sets of pieces. The aim is to swap the positions of three male frogs and three female frogs.

You can try to solve the puzzle using three boys, three girls, and seven chairs. Alternatively, you can use three counters of one color to represent the female frogs and three counters of another color to represent the male frogs. Then place them on a board like this so that the circles represent lilypads in a pond.

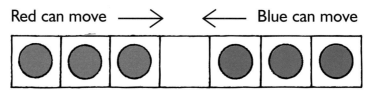

There are two ways that a frog can move:
• *Either* slide one space to an empty lilypad, in the direction shown.
• *Or* jump over one frog of the opposite sex to an empty pad beyond.

FROGS CHALLENGE 1

It is possible to swap the positions of the two sets of frogs in 15 moves. Can you do it?

FROGS CHALLENGE 2

Make a nine-lilypad pond with four male frogs and four female frogs. Can you solve this game of "Frogs"?

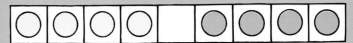

COUNTER CHALLENGE

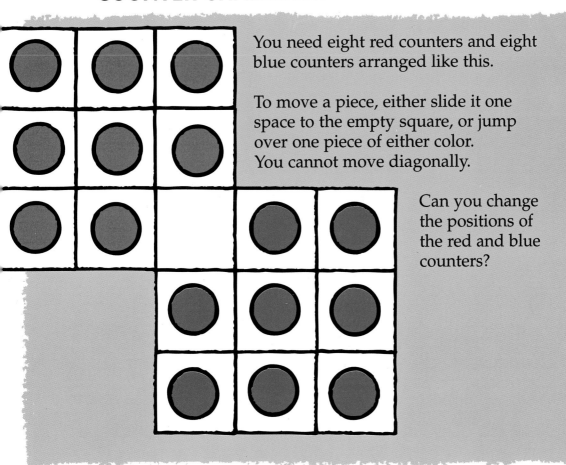

You need eight red counters and eight blue counters arranged like this.

To move a piece, either slide it one space to the empty square, or jump over one piece of either color.
You cannot move diagonally.

Can you change the positions of the red and blue counters?

SWAPPING PUZZLES

In each of these swapping position puzzles, you may only slide pieces horizontally or vertically. No jumping is allowed.
Can you solve the first puzzle in 16 moves and the second in 24 moves?

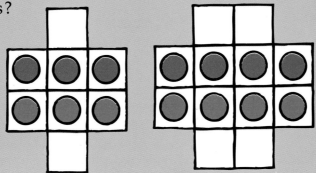

TANGRAMS

The tangram is an ancient Chinese puzzle. It is like a jig-saw puzzle in that it fits together a set of shapes. There are seven tangram pieces in all that can be cut from a square.

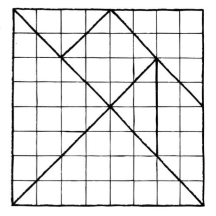

MAKE A TANGRAM

Make your own tangram by drawing the outline on some graph paper.

TANGRAM PICTURES

All of these pictures have been made by arranging the seven pieces. Try to make them.

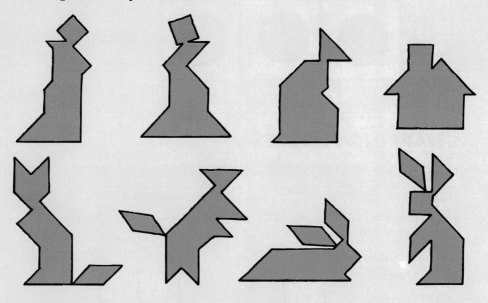

In this picture "Away in a Manger," each part is made from the same separate tangram. Create your own picture from tangrams.

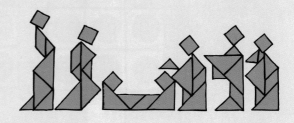

Notice the shape of each of the tangram pieces. There is one square, one **parallelogram** and five **isosceles right triangles** (two small, one medium-sized, and two large).

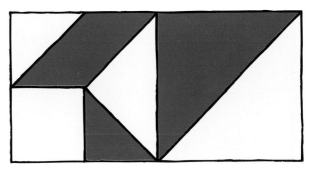

As well as arranging the pieces to create pictures, they can be arranged to make geometrical shapes. For example, this shows how to arrange them to make a rectangle.

TANGRAM SHAPES

Can you arrange the seven tangram pieces above to make:
• a parallelogram
• a **trapezoid**
• a right triangle?

TANGRAM PUZZLE

Make this 3-piece tangram puzzle by drawing a square on cardboard.
Mark the mid-points 1 and 2.
Use these as guides to draw the two lines.
Cut along the two lines to make three pieces.
Label them A, B, and C.

Use the three pieces to see how many of these you can make:
• a triangle using A and B
• a triangle using A, B, and C
• a rectangle using A, B, and C
• a trapezoid using A and B
• a trapezoid using A and C
• a trapezoid using B and C
• a trapezoid using A, B, and C
• a **pentagon** using A, B, and C.

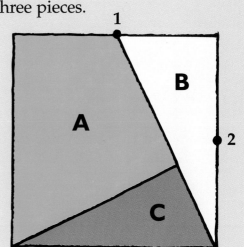

GLOSSARY

diagonal
A line drawn from one corner to an opposite corner.

horizontal lines
Straight lines drawn from left to right.

isosceles triangle
A triangle that has two equal sides.

magic number
The total of the rows, columns, and diagonals in a magic square.

magic square
A square grid of numbers in which all rows, columns, and diagonals have the same total.

multiple
The multiples of a number are those that can be evenly divided by that number. For example, the multiples of 9 are 9, 18, 27, and so on.

palindromic number
A number that reads the same forward and backward.

parallelogram
A quadrilateral with two pairs of parallel sides.

pentagon
A polygon with five sides.

pentagram
A shape made by joining the diagonals of a regular pentagon.

right triangle
A triangle that has one right angle.

symmetrical pattern
Patterns are symmetrical when they have two matching halves, one a reflection of the other.

trapezoid
A quadrilateral with one pair of parallel sides.

vertical lines
Straight lines drawn from top to bottom, at right angles to a horizontal line.

INDEX

ANSWERS

p. 7

Coin Sharing
1 penny, 2 nickels, 2 dimes, 1 quarter *and* 1 penny,
2 nickels, 2 dimes, 1 quarter
1 penny, 4 nickels, 1 dime, 1 quarter *and* 1 penny,
3 dimes, 1 quarter
1 penny, 1 nickel, 2 quarters *and* 1 penny, 3 nickels,
4 dimes

p. 8

Straw Squares

Straw Division

p. 9

Straw Equations
I=III−II   V+IV = IX

p. 10

Triplets
2,3,7; 6,5,1; 6,4,2; 7,4,1; 8,3,1; 9,2,1; 5,4,3.
6,5,4; 6,7,2; 6,8,1; 4,8,3; 4,9,2; 4,10,1; 5,8,2;
5,9,1; 5,7,3; 10,3,2.

p. 11

Sixteen Card Puzzle
One possible set of solutions is:

2	4	1	3		1	2	3	4		3	2	1	4
3	1	2	4		3	4	1	2		1	4	3	2
1	3	4	2		4	1	2	3		4	1	2	3
3	2	4	1		2	3	4	1		2	3	4	1

p. 12

Missing Dots

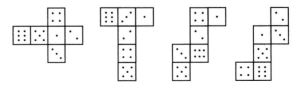

46

p. 13   Impossible Number Cubes

p. 19   **Number Maze**
Total of 15: pathway +6, +3, +5, +1
Largest total, 18; smallest total, 13
Different totals: 13, 14, 15, 16, 17, 18

p. 26   **Palindromic numbers are:**
101, 111, 121, 131, 141, 151, 161, 171, 181, 191
202, 212, 222, 232, 242, 252, 262, 272, 282, 292
303, 313, 323, 333, 343, 353, 363, 373, 383, 393

p. 30   **Magic Number - 65**